Y0-DPI-776

DOE/NV/10162-9

GEOLOGICAL RECONNAISSANCE AND CHRONOLOGIC STUDIES

by

Jonathan Davis[1]

Social Sciences Center
Desert Research Institute
University of Nevada System
Reno, Nevada

March, 1983

[1]Assistant Research Professor, Social Sciences Center

The work upon which this report is based was supported in part by the U.S. Department of Energy under Contract DOE-AC08-81NV10162.

ISBN 0-945920- 33 - 4

ISSN 0897-6376

ABSTRACT

There are several possible scenarios by which a radioactive waste storage facility in the unsaturated zone could be compromised; among them erosion, water table rise, or downward percolation of water. In order to assess these risks, the geologic and climatic events of the past few million years can be used to project the future of the unsaturated deposits. Geologic reconnaissance on and around the NTS was undertaken to identify specific evidence of depositional, erosional, and hydrologic events, as well as to develop an understanding of the timing of these events.

Several kinds of evidence were noted and studied: layers of voclanic ash in the basin-fill sediments were discovered and dated at 11 to 5 million years old, showing that the modern valleys and ranges are at least 11 million years old. Exposure of these ash layers by erosion has taken 5 million years, implying that additional millions of years must pass before modern closed basins on the NTS are eroded. Detailed study of young sediments in Las Vegas Valley suggest that water tables stood at 926 m as recently as 14,000 years ago. To the northeast of the NTS, sediments in basin bottoms also reflect high water tables until about 7,000 years ago, but sediments on the NTS proper do not show this effect during the last 700,000 years. The observed relation between erosion due to downwearing of mountain ranges and infilling of valleys suggests that as these processes continue, only the uppermost parts of present alluvial fans will become eroded.

Of these observations and conclusions, only the inferred 926 m water table stand in Las Vegas Valley suggests possible future adverse impact to a radioactive waste facility in the unsaturated zone of the NTS.

CONTENTS

TABLES

FIGURES

INTRODUCTION

Although the advantages of disposal of radioactive waste in the unsaturated zone are well recognized (Winograd, 1980), certain possible flaws in this disposal scheme have been identified which rest upon possible changes in geologic or hydrologic conditions. Winograd and Doty (1980) have discussed some of these, and have shown why a knowledge of past water table conditions is essential to critical evaluation of the NTS as a waste repository. Some of the possible geologic/hydrologic changes which could compromise a repository include:

- a rise in water table, which at best would reduce the useful volume of the unsaturated zone, and at worst would saturate disposed waste.
- precipitation and concommitant infiltration could increase such that downward percolation through the unsaturated zone ensued.
- erosion could expose or even disperse the waste on the surface.
- localized discharge from the mountains could percolate downward through a repository.
- faulting or tilting could change the hydrologic situation.

As discussed by Winograd and Doty (1980), one of the most powerful means of assessing the likelihood of future geologic events, like those listed above, is to investigate the geologic record of the past few millions of years, during which substantial changes in climate have occurred (Mehringer, 1967; 1977), as well as various changes in depositional regime (Haynes, 1967; Quade, 1983). It is not unreasonable to assume that the cyclic changes of climate during the geologically recent past represent the extremes of climatic changes likely in the future, and that processes which can be seen to be progressive (i.e., basin filling) will continue as they have before.

PREVIOUS RESEARCH

Although certain studies have been undertaken recently to decipher the geologic record and stratigraphy of the last few million years of geologic time on the NTS, (e.g., Hoover and others, 1981) very few studies exist where sediments and associated paleoclimate or paleoenvironmental data have been presented in a dated context. Haynes (1967) described the record of the last 50,000 years or so as preserved from Tule Springs to Corn Creek Springs, north of Las Vegas (this work has been partially reinterpreted by Quade (1983), discussed later in this report), where sediments were laid down by springs and small streams. Pollen and other paleoenvironmental evidence from Tule Springs were discussed by Mehringer (1967), and Mehringer and Warren (1976) presented evidence of environmental change in Ash Meadows during the last 5,000 years.

Study of plant macrofossils from pack rat middens has proved very useful in deciphering the past vegetation of southern Nevada (Mehringer and Ferguson, 1969; Spaulding, 1977; Van Devender and Spaulding, 1979), and pack rat midden studies are underway in several areas adjacent to the NTS. However, vegetation reflects almost entirely the soil conditions and microclimate of the immediate area, and cannot directly show the location of water tables, for instance, as well as evidenc left in the sediment itself. Unfortunately, in southern Nevada comparatively little is known about the sediments themselves and their age.

In northern Nevada, however, a fairly well dated geologic sequence for the past few tens of thousands of years exists. Studies by Morrison (1964), who described stratigraphic units, followed by Benson (1978) and Davis (1978; in press) who provided radiometric chronology and time-stratigraphic marker horizons, have resulted in the development of an understanding of climatic and environmental changes through the last 35,000 years (Davis, 1982). In

2

particular, Morrison described the mid-Holocene Toyeh Soil, about 5,000 years old, which seems to be present over vast areas of Nevada and the western United States (Haynes, 1968; Davis, 1978) and serves as a time marker which divides the last 5,000 years from earlier geologic time. Other particularly useful time markers are the Mazama volcanic ash layer, eruptd from Crater Lake, Oregon, about 6,800 years ago, which occurs throughout northern and east-central Nevada, and volcanic ash layers from Mono Craters, which are less than 2,000 years old, and occur as far east and south as central Nevada (Davis, unpublished data). It has been an important part of this project to attempt to trace some or all of these markers from the areas where they are recognized into the NTS or its vicinity.

SCOPE OF INVESTIGATIONS

Presented below are preliminary results of field reconnaissance conducted over 1980, 1981, and 1982, on and around the NTS. The area investigated by Winograd and Doty (1980; Ash Meadows ground water basin) was not closely examined to avoid duplication of their effort. Reconnaissance was oriented towards developing an overview of late Cenozoic events in the region, and identification of localities and lines of inquiry which would address the possible changes in geologic context which would compromise, or perhaps enhance, the integrity of a radioactive waste repository on the NTS. Specifically, the following were searched for:

- evidence of past water-table stands (carbonate deposits)
- exposure of stratigraphic records of the late Cenozoic
- means of correlation within the late Cenozoic (volcanic ash layers)

3

- evidence of patterns of deposition and erosion
- exposures of soil profiles, showing the changes in morphology through time caused by soil horizonation.

Two lines of inquiry developed from this investigation have been developed into monographs written by Metcalf (1982) and Quade (1983), although Quade was not supported by this project.

Field reconnaissance was conducted in the following areas: on the NTS particularly in Frenchman Flat and along Fortymile and Yucca washes; from Tule Springs to Indian Springs in Las Vegas Valley; from Corn Creek Springs to Alamo through Desert National Wildlife Range; Coyote Springs Valley and Pahranagat Valley; White River Valley; along Meadow Valley Wash, Garden Valley; Cave Valley; southern Steptoe Valley; Reveille Valley; Stone Cabin Valley; Oasis Valley, Crater Flat, Stewart Valley, Chicago Valley, Pahrump Valley, and Ash Meadows. Some of these locations are shown in Figure 1.

PRELIMINARY RESULTS

Results of the geological studies will be presented under several headings: the chronological results of Metcalf (1982), which pertain to the age of faulting and the rate of erosion; the work of Quade (1983) which has interesting implications as to the history of water table fluctuations in the NTS region; and unpublished observations by Davis, Clerico and Quade which also relate to water table fluctuations, and to depositional history of the NTS. Much of the NTS work is still in progress and will be presented in fuller form in future reports.

VOLCANIC ASH CHRONOLOGY

Metcalf (1983) studied several layers of volcanic ash (or tephra) which crop out in the uppermost basin-filling sediments south and east of the NTS. Using the Potassium-

LEGEND FOR FIGURE 1.

     1: Tule Springs
     2: Corn Creek Springs
     3: Indian Springs
     4: Alamo
     5: Coyote Springs Valley
     6: Pahranagat Valley
     7: White River Valley
     8: Meadow Valley Wash
     9: Garden Valley
    10: Cave Valley
    11: Steptoe Valley
    12: Reveille Valley
    13: Stone Cabin Valley
    14: Oasis Valley
    15: Stewart Valley
    16: Chicago Valley
    17: Pahrump Valley
    18: Ash Meadows
    19: Crater Flat
    20: Frenchman Flat
    21: Yucca and Fortymile Washes
    22: Las Vegas Valley

* Radioactive Waste Management Site

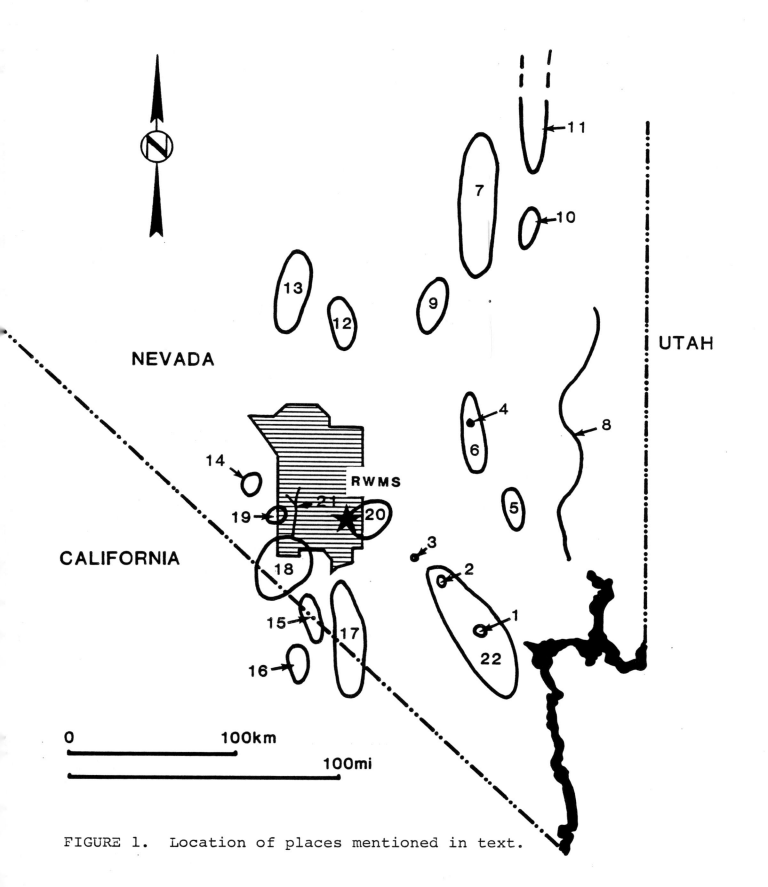

FIGURE 1. Location of places mentioned in text.

Argon dating method, she determined that the volcanic ash layers, and the sediments which contain them, are from about 5 million to about 11 million years old. These sediments lie in the present valleys and have been cut into and exposed by headward erosion from the Colorado and Amargosa rivers. From these relations, certain conclusions can be drawn:

- that the present ranges and valleys were in existence by about 11 million years ago, and faulting and tilting since have not been sufficient to significantly disrupt the basin fills.

- that by 5 million years ago, headward erosion by the Colorado and Amargosa rivers had not yet breached the basins where the ashes were discovered; this headward erosion has occurred since then.

It would be tempting to say also that very little deposition has occurred during the last 5 million years, but this may be true only in the basins which have been breached by headward erosion, and in these basins the sediment may have simply bypassed the sites studied and been carried downstream. During the same period of time, considerable amounts of sediments may have accumulated in the closed basins to the north.

Reconnaissance by Davis, Clerico, and Quade, described below, suggests that deposition is a continuing process in each of the closed basins not yet breached by the headwaters of the Colorado or the Amargosa; the areas in these valleys below any fan head trenches will continue to receive sediments until the basins are breached. It is difficult to estimate when, if ever, such breaching will occur but Frenchman Flat, for instance, is separated from the Amargosa River drainage by only a single ridge. Adjacent Mercury Valley drains through Point of Rocks into the Amargosa Desert, and once the ridge is eroded away, Frenchman Flat in turn will drain and dissection of the basin fill there can begin. However, Metcalf's results indicate that such headward erosion requires millions of years, and it may be estimated that a period of time on the order of five million years will

be required before significant erosion of Frenchman Flat is likely to occur.

WATER TABLE HISTORY IN LAS VEGAS VALLEY

Winograd and Doty (1980) studied the water table history of the Frenchman Flat-Ash Meadows area, and concluded that the likely magnitude of water table rise during the Pleistocene was about 30 m, up to an elevation of about 760 m above sea level, or 180 m beneath the surface of Frenchman Flat. This study was confined to the area of the NTS and Ash Meadows because Winograd and Doty regarded this as an isolated ground-water basin. Recent geologic work adjacent to this basin by Quade (1983), which was partially supported by DRI and DOE, may suggest that Pleistocene water table levels were much higher than this.

Quade studied the sediments of the area from Indian Springs to Tule Springs State Park, in northern Las Vegas Valley, in an effort to determine their age and genesis. Earlier, Haynes (1967) had studied equivalent sediments east of Tule Springs State Park, concluding that a pluvial lake had existed in Las Vegas Valley. Quade modified the interpretations of Haynes, inferring that very shallow, marshy, standing water had occupied the axis of northern Las Vegas Valley during the time from about 30,000 to about 14,000 years ago. Further, Quade recognized that an horizon of calcareous nodules in the sediments which crosscuts stratification reflected the existence of a water table and associated capillary fringe which stood as high as 926 m above sea level in the Corn Creek Springs area, and provided the source of the water in the marsh in the valley axis.

This conclusion does not necessarily mean that Pleistocene water tables below Frenchman Flat (which lies at 940 m) were only 14 m below the surface of the flat, because Corn Creek Springs is regarded as being in a separate ground water basin and because water tables normally slope away from recharge areas. The Corn Creek area is on the flanks of the

Spring Mountains, a major recharge area. However, Quade's results suggest that Pleistocene water tables were at the surface in Las Vegas Valley at an elevation only slightly lower than that of Frenchman Flat, and therefore the subject is deserving of further investigation. Winograd and Doty (1980) suggested that careful study of a continuous core from Frenchman Flat would be valuable in this regard.

DEPOSITIONAL HISTORY ON AND NORTHEAST OF NTS

Davis and Clerico (unpublished data) have conducted geologic reconnaissance on the NTS and in the surrounding area especially to the northeast, in an attempt to identify means of dating the basin fill sediments, such as volcanic ash layers or soil-stratigraphic markers, and to find areas where erosional exposure provides access to the geological record of the basin fill. These efforts have resulted in the delineation of a regional pattern in sedimentation during the recent geologic past, recognition of a consistent, repeated sequence of depositional and soil-forming events in the basins northeast of the NTS, and description of a series of depositional and erosional events on the NTS itself.

The Mt. Mazama volcanic ash time marker was erupted from Crater Lake, Oregon, about 6800 years ago; its chemical and physical properties allow it to be distinguished unambiguously throughout much of the northwestern United States (Kittleman, 1973; Davis, 1978). The Mazama ash is known from various localities in central Nevada, and in this study, new localities have been identified in Stone Cabin, Hot Creek, Railroad, and White River Valley, or along the northern border of the Nellis Range (Davis, unpublished data; Table 1). However, the Mazama ash has not been found in Cave and Steptoe valleys, along Patterson Wash, Meadow Valley Wash, or in Garden or Reveille valleys. Nonetheless, the ash, where found, lies in a characteristic stratigraphic position.

In Hot Creek Valley in particular, the Mazama ash lies upon as much as 3 m of strikingly dark, organic-rich silt

TABLE 1. Electron microprobe chemistry of glass from Mazama ash specimens from area north and east of NTS, in per cent, recalculated water-free.

	SiO_2	Al_2O_3	Fe_2O_3	MgO	MnO	CaO	BaO	TiO_2	Na_2O	K_2O	Cl
Known Mazama ash from Crater Lake											
CB-39	72.4	14.6	2.25	.48	.05	1.60	.07	.44	5.2	2.7	.18
Jakes Wash, White River Valley											
GS-94	73.2	14.6	2.22	.41	.04	1.56	.07	.45	4.7	2.6	.17
DR-93	72.6	14.4	2.25	.48	.05	1.68	.09	.41	5.2	2.7	.17
Moores Station Wash, Hot Creek Valley											
DR-84	72.6	14.7	2.23	.47	.06	1.66	.09	.43	4.9	2.7	.20
Twin Springs Slough, Railroad Valley											
LM-42	72.3	14.7	2.20	.45	.05	1.61	.10	.46	5.3	2.6	.15
Stone Cabin Ranch, Stone Cabin Valley											
DR-73	72.4	14.8	2.14	.46	.05	1.58	.09	.42	5.3	2.6	.16
DR-74	72.6	14.6	2.19	.46	.04	1.64	.08	.41	5.1	2.7	.16

which is finely laminated and seems to reflect wet meadow, high water-table conditions prior to 6800 years ago. Over-lying the ash are up to 2 m of vaguely stratified tan sand, which is similar to the sediment deposited by overbank flow along washes in the area today. Farther to the south, in Garden Valley and along Patterson Wash, although the ash marker is not present, the stratigraphic sequence is the same, and the implication is that during the latest Pleisto-cene, local water tables in the valley bottoms were at the surface and supported abundant wet meadow vegetation, but that these conditions ceased by about 6800 years ago, and more arid conditions have prevailed since.

In attempting to trace this sequence onto the NTS, how-ever, difficulties are encountered. On the NTS, neither the organic-rich silt nor the tan sand can be identified, and the general character of deposition during the last ten thousand years has been quite different. On the NTS, for instance, along Yucca and Fortymile washes, the Quaternary alluvium is almost entirely gravel, and the stratigraphic sequence defined to the northeast cannot be recognized. This change seems to be a major break in sedimentation; the entire region south and west of the NTS to the Sierra Nevada is characterized by gravelly alluvium, rather than silt or sand. Exposures along Yucca and Fortymile washes on the western NTS include vol-canic ash similar to the 700,000 year old Bishop ash (Davis, unpublished data; Table 2) in the gravel sequence, which implies that the gravelly alluvium has been deposited in the NTS predominately for the last 700,000 years. (The Bishop ash was identified in eolian deposits in NTS by Hoover and others, 1981.) It is probable that the gravel reflects tor-rential, flashy stream discharge produced in response to sudden rainfall events, whereas the finer sediments to the northeast indicate that precipitation is gentler and less torrential, including gradual snowmelt. The major implica-tion is that the modern flashy pattern of discharge on the NTS has been typical of the area for at least 700,000 years,

TABLE 2. Electron micropbobe chemistry of glass from Bishop ash samples from known localities (Sarna and others, 1980) and from Yucca Wash. In per cent, recalculated water-free. "ND" means "not determined".

	SiO	Al_2O_3	Fe_2O_3	MgO	MnO	CaO	BaO	TiO_2	Na_2O	K_2O	Cl
Known Bishop ash specimens, sample numbers from Sarna and others (1980)											
8	77.4	12.7	.75	.04	.04	.45	ND	.06	3.3	5.2	ND
10	77.3	12.8	.78	.03	.03	.46	ND	.05	4.1	4.3	.06
12	77.3	12.8	.79	.04	.05	.47	ND	.05	4.6	3.8	.07
Ash from alluvium along Yucca Wash											
DR-66	77.7	12.7	.76	.03	.04	.44	.004	.07	3.8	4.5	.08

and that the high water table conditions which occurred before 6800 years ago in the area to the northeast have not
occurred on the NTS.

DEPOSITIONAL AND EROSIONAL PATTERNS ON ALLUVIAL FANS

Because the bulk of the unsaturated alluvium on the NTS
comprises sediments of alluvial fans and bajadas, special
attention has been devoted to patterns of deposition and
erosion on these landforms. It is possible to determine the
relative ages of parts of alluvial fans by mapping from
aerial photographs, by examination of surface features such
as desert pavement and desert varnish (Hoover and others,
1981), and by examination of soil profile development, which
in the NTS area comprises mostly carbonate accumulation.

Although alluvial fans are fundamentally depositional
features, downwearing of their mountain sources can lead to
entrenchment of the upper parts of the fans even though base
level in the basin is rising due to accumulation of trans-
ported sediments. Such fan head trenches are common in and
around the NTS. Ideally, a waste disposal site should lie
lower on the fan or bajada than the zone of present and
future entrenchment, yet no lower than necessary, so that
deposition will not endanger the disposal operation (flash
flooding), and percolation associated with deposition will be
unlikely. Furthermore, the higher on the fan the site is
located, the thicker the unsaturated sediment beneath it will
be.

General observations of the NTS region, as well as
detailed examination and mapping of the northwest part of
Frenchman Flat, shows that fan head entrenchment has occurred
only at the extreme heads of fans, usually in embayed
mountain fronts. Below the embayments, recent sediments
mantle the fans and older sediments are buried. In the
northwestern part of Frenchman Flat, the same pattern is
observed; entrenchment of fan deposits has occurred only in
the embayment north and west of the disposal site, and

12

mapping shows that the present disposal site is located on the most recent sediments. Therefore, the present disposal site seems well located with respect to erosional and depositional balance on the fans.

PRELIMINARY SUMMARY

Of the lines of inquiry discussed above, only Quade's inference of a 926 m water table in northwestern Las Vegas Valley is a cause for concern with respect to disposal of radioactive waste in the unsaturated zone of the NTS. Davis and Clerico have found evidence of similar high water tables during the late Pleistocene, but farther to the east and north; this evidence does not involve ground water basins adjacent to the NTS, and Davis, Clerico and Quade's investigation of the NTS proper suggests that water tables there have not reached the surface during the last 700,000 years. Metcalf's research supports the idea that the NTS is a comparatively stable area erosionally and tectonically, and suggests that breaching of NTS basin fills by headward erosion is millions of years in the future.

REFERENCES

Benson, Larry V., 1978. Fluctuations in the level of pluvial
    Lake Lahontan during the last 40,000 years. _Quaternary
    Research_ 9: 300-318.

Davis, J.O., 1978. Quaternary tephrochronology of the Lake
    Lahontan area, Nevada and California. Nevada
    Archeological Survey, University of Nevada. _Research
    Paper_ No. 7, Reno.

Davis, J.O., 1982. Bits and pieces: the last 35,000 years in
    the Lahontan area, in Madsen, D.B., and O'Connell, J.F.
    (editors) _Man and Environment in the Great Basin_, So-
    ciety for American Archaeology, Papers, No. 2, p. 53-
    75.

Davis, J.O., in press. Level of Lake Lahontan during depo-
    sition of the Trego Hot Springs tephra about 23,400
    years ago. _Quaternary Research_, V. 19, No. 2.

Haynes, C.V., Jr., 1967. Quaternary geology of the Tule
    Springs area, Clark County, Nevada, in Wormington, H.M.
    and Ellis, Dorothy (editors), _Pleistocene Studies in
    Southern Nevada_, Nevada State Museum Anthropological
    Papers, No. 13, p. 15-104.

Haynes, C.V., Jr., 1968. Geochronology of late-Quaternary
    alluvium, in Morrison, R.B., and Wright, H.E., Jr.
    (editors) _Means of Correlation of Quaternary Succes-
    sions_, International Association for Quaternary Re-
    search, VII Congress, Proceedings, v. 8, p. 591-631.

Hoover, D.L., Swadley, W.C., and Gordon, A.J., 1981. Cor-
    relation characteristics of surficial deposits with a
    description of surficial stratigraphy in the Nevada Test
    Site region. U.S. Geological Survey _Open-File Report_
    81-512, 44 p.

Kittleman, L.R., 1973. Mineralogy, correlation, and grain-
    size distribution of Mazama tephra and other postglacial
    pyroclastic layers, Pacific Northwest. _Geological So-
    ciety of America Bulletin_ 84: 2957-2980.

Mehringer, P.J., Jr., 1967. Pollen analysis of the Tule
    Springs Area, Nevada. H.M. Wormington and Dorothy Ellis
    (editors) _Nevada State Museum Anthropological Papers_ No.
    13, pp. 129-200.

Mehringer, P.J., Jr., 1977. Great Basin Late Quaternary Environments and Chronology. In Don D. Fowler (editor) Models and Great Basin Prehistory: A symposium. Nevada University, Desert Research Institute, Publications in the Social Sciences No. 12, Reno and Las Vegas, p. 113-167.

Mehringer, P.J., Jr., and Ferguson, C.W., 1969. Pluvial occurence of Bristlecone Pine (Pinus aristata) in a Mohave Desert Mountain Range. Journal of the Arizona Academy of Sciences, v. 5, p. 284-292.

Mehringer, P.J., Jr., and Warren, C.N., 1976. Marsh, dune, and archaeological chronology, Ash Meadows, Amargosa Desert, Nevada. Elston, R. (editor) Holocene environmental change in the Great Basin, Nevada University, Archaeological Survey, Research Paper 6, p. 120-150.

Metcalf, L.A., 1982. Tephrostratigraphy and Potassium-Argon Age Determinations of Seven Volcanic Ash Layers in the Muddy Creek Formation of Southern Nevada. Nevada University, Desert Research Institute, Water Resources Center, technical report prepared for U.S. Dept. of Energy under contract DE-AC08-80NV10162, 187 p.

Morrison, R.B., 1964. Lake Lahontan: geology of southern Carson Desert, Nevada. U.S. Geological Survey Professional Paper 401.

Quade, J., 1983. Quaternary geology of the Corn Creek Springs area, Clark County, Nevada. unpublished M.S. Thesis, Geosciences, University of Arizona, Tucson, 135 p.

Sarna-Wojcicki, A.M., Bowman, H.R., Meyer, C.E., Russell, P. C., Asaro, F., Michael, H., Rowe, J.J., Baedecker, P.A., and McCoy, G., 1980. Chemical analyses, correlations, and ages of late Cenozoic tephra units of east-central and southern California. U.S. Geological Survey Open File Report 80-231, 52 p.

Spaulding, W.G., 1977. Late Quaternary vegetational change in the Sheep Range, southern Nevada. Journal of the Arizona Academy of Sciences v. 12, p. 3-8.

Van Devender, T.R., and Spaulding, W.G., 1979. Development of vegetation and climate in the southwestern United States. Science v. 204, p. 701-710.

Winograd, I.J., 1980. Radioactive waste storage in thick
     unsaturated zones. Science, v. 212, no. 4502, p. 1457-
     1464.

Winograd, I.J., and Doty, G.C., 1980. Paleohydrology of the
     southern Great Basin, with special reference to water
     table fluctuations beneath the Nevada Test Site during
     the late (?) Pleistocene. U.S. Geological Survey, Open
     File Report 80-569, 91p.

DOE DISTRIBUTION LIST AND ADDRESSES   (R1/83)

James Morley
USDOE
Albuquerque Operations Office
P.O. Box 5400
Albuquerque, NM 87115

R. D. Duncan, c/o D. A. Nowack
USDOE
Nevada Operations Office
P.O. Box 14100
Las Vegas, NV 89114

D. A. Nowack
USDOE
P.O. Box 14100
Las Vegas, NV 89114

R. M. Nelson
USDOE
Nevada Operations Office
P.O. Box 14100
Las Vegas, NV 89114

D. L. Vieth
USDOE
Nevada Operations Office
P.O. Box 14100
Las Vegas, NV 89114

J. H. Dryden
USDOE
Nevada Test Site Support Office
Mercury, NV 89023

CETO Cibrary
USDOE
Mercury, NV 89023

H. Hollister
USDOE
Forrestal Building
1000 Independence Ave., SW
Washington, DC 20585

W. W. Hoover
Office of Military Application
USDOE, GTN

J. E. Baublitz
Director of Rem. Act. Project
(NE-24) GTN
USDOE

Office of Public Affairs
Robert C. Odle, Jr.
USDOE
Forrestal Building
1000 Independence Ave. SW
Washington, DC 20585

D. E. Patterson
USDOE, GTN

J. R. Maher
USDOE, GTN

M. G. White
USDOE, GTN

USDOE
Technical Information Center
P.O. Box 62
Oak Ridge, TN 37830

Defense Nuclear Agency
Test Construction Division
FCTC/DNA
ATTN: J. W. LaComb, C. Snow
Mercury, NV 89023

Defense Nuclear Agency
Director
ATTN: SPSS, Milton Peak, Eugene
   Sevin, T.E. Kennedy
SPTD
Washington, DC 20305

C.E. Keller
FCTMC/DNA
Kirtland AFB, NM 87115

Library
Los Alamos National Scientific
   Laboratory
P.O. Box 1663
Los Alamos, NM 87545

N.W. Howard
Lawrence Livermore National
   Laboratory
P.O. Box 808
Livermore, CA 94550

D.L. Springer
Lawrence Livermore National
   Laboratory
P.O. Box 808
Livermore, CA 94550

L. McHague
Lawrence Livermore National
   Laboratory
P.O. Box 808
Livermore, CA 94550

Technical Information Library
   Division
Lawrence Livermore National
   Laboratory
P.O. Box 808
Livermore, CA 94550

John R. Duray
The Bendix Corporation
Field Engineering Corp
Redland
Grand Junction, CO 81501

Library
U.S. Army Corps of Engineers
Waterways Experiment Station
Vicksburg, MS 39180

P.L. Russell
U.S. Bureau of Mines
Denver Federal Center
Building 20
Denver, CO 80225

Geologic Data Center
U.S. Geological Survey
Mercury, NV 89023

Library
U.S. Geological Survey
Box 25046, Mail Stop 954
Denver, CO 80225

Library
U.S. Geological Survey
Middlefield Road
Menlo Park, CA 94025

Chief Hydrologist, WRD
ATTN: Radiohydrology Section
U.S. Geological Survey
USGS National Center
12201 Sunrise Valley Dr.
Reston, VA 22092

Library
U.S. Geological Survey
USGS National Center
12201 Sunrise Valley Dr.
Reston, VA 22092

Military Geology Unit
U.S. Geological Survey
USGS National Center
12201 Sunrise Valley Dr.
Reston, VA 22092

J.C. Reed, Jr.
U.S. Geological Survey
USGS National Center
12201 Sunrise Valley Dr.
Reston, VA 22092

W.B. McKinnis
Lawrence Livermore Laboratory
Mercury, NV 89023

B.G. Edwards
Sandia Laboratories
Mercury, NV 89023

Howard P. Stephens
Sandia Laboratories
Mercury, NV 89023

R. Lincoln
Sandia Laboratories
Mercury, NV 89023

G. H. Heilmeier
Defense Advanced Research
   Projects Agency
1400 Wilson Blvd.
Arlington, VA 22209

Paul R. Fenske
Desert Research Institute
Water Resources Center
Reno, NV 89512

Water Division
Environmental Projtection Agency
1200 6th
Seattle, WA 98101

G. E. Schweitzer
Environmental Protection Agency
Environmental Monitoring
   Systems Laboratory
944 E. Harmon Ave.
Las Vegas, NV 89109

H. F. Mueller
National Oceanic and Atmospheric
   Administration
Weather Research Center
P.O. Box 14985
Las Vegas, NV 89114

Grant Bruesch
Fenix & Scisson, Inc.
Mercury, NV 89023

A. E. Gurrola
Holmes & Narver, Inc.
2753 S. Highland Dr.
Las Vegas, NV 89109

Resident Manager
Holmes & Narver, Inc.
Mercury, NV 89023

E. M. Romney
Laboratory of Nuclear Medicine
University of California at
   Los Angeles
900 Veteran Ave.
Los Angeles, CA 90024

William E. Wilson, Chief
Nuclear Hydrology Program
U. S. Geology Survey
Denver Federal Center
M/S-416
P.O. Box 25046
Denver, CO 80225

Henry L. Melancon
USDOE
P.O. Box 14100
Las Vegas, NV 89114

T. M. Humphrey, Jr.
USDOE
P.O. Box 14100
Las Vegas, NV 89114

Mitchell P. Kunich
USDOE
P.O. Box 14100
Las Vegas,NV 89114